PLUMBING

By Whitney Hoovler and Jeffrey Pulis

CASTLE BOOKS

Distributed by
BOOK SALES, INC.
110 Enterprise Avenue
Secaucus, N.J. 07094

Our staff has done its best to make this text accurate, easy to use and clear. However, we are not responsible in any way, if in using this book, you should make any errors.

A SUNRISE BOOK

Copyright © 1975 by Sunrise Books, New York, N.Y.
Printed in the United States of America.

PLUMBING

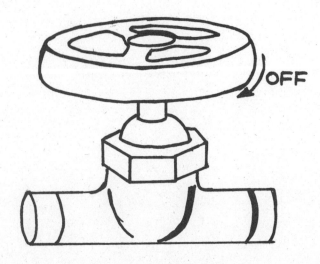

AN OUNCE OF PREVENTION

Going on summer vacation? Turn off switch on the oil burner. Otherwise hot water will be produced unnecessarily for an empty house.

Going on an off-season vacation? Is there any chance of freezing weather? Keep the heat on, setting the thermostat at 60 degrees to prevent freezing of pipes.

Turn off the water to the washing machine whenever it is not in use. If you don't, water pressure on the hoses will eventually make the hoses give way, causing a flood.

There's a lot of talk lately of putting a brick in the toilet tank to save water. Of course you can accomplish the same thing by adjusting the float. But remember, the purpose of the water is sanitation. You can use less but you may find that daily cleaning of the bowl is necessary.

There is a gadget now sold that reputedly dissolves hair in drains. If hair is a problem at your house, you may wish to look into it.

TABLE OF CONTENTS

INTRODUCTION
LEAKY FAUCETS — 3
  Globe Faucets
    Leak From Under Handle — 14
    Leak From End Of Spout — 16
  Basin Or Tub Faucets
    Leak From Under Handle — 20
    Leak From End Of Spout — 22
  Ball Faucets, Single Handle
    Leak From Under Handle — 26
    Leak From End Of Spout — 28
    Leak Top Or Bottom Of Spout — 30
    Cleaning Clogged Diverter Assembly — 32
  Deck Faucets
    Replacing Basin Or Kitchen Faucets — 34

TOILETS
  Maintenance — 36
  Tank Leaks
    Replacing Flush Ball — 38
    Repairing Float Apparatus — 40
    Repairing Float Valve — 42

FLUSHING PROBLEMS — 44
  Toilet Bowls
    Leak Between Tank & Bowl — 46
    Leak Between Bowl & Floor — 48

TABLE OF CONTENTS

HEATING SYSTEMS	51
Sweating Pipes	54
Water Hammer	55
Prevent A Leak	
Maintenance	
Steam Heat - Yearly Clean Out	56
Steam Heat - Banging Radiators	58
Hot Water Heat - Banging Radiators	61
WATER SUPPLY SYSTEM	65
Minor Problems	
Recognizing Pipe Materials	66
Leaky Pipes	
Basics	69
Leak At Connection, Plastic or Polyethylene	70
Changing to Removable Coupling, Plastic Pipe	72
Removing Soldered Pipe From Fitting, Metal Pipes	74
Working With Pipe	
Cutting Pipe	77
Threading Pipe	79
Soldering Pipe	81
Bending Pipe	84
DRAINAGE SYSTEM	86
Drain Pipes	87
Blockage	88
OUTDOOR PIPES	91
Thawing A Frozen Pipe	92
INSTALLING A SUMP PUMP.	94

PLUMBING

BASIC TOOLS

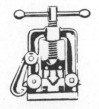
PIPE VISE

SCREWDRIVER

CLOSET AUGER

COLD CHISEL

MONKEY WRENCH

PIPE THREADER

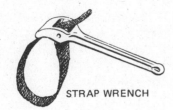
STRAP WRENCH

ADJUSTABLE PIPE WRENCH

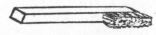
WIRE BRUSH

PLUMBING

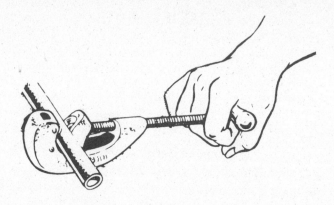

BASIC TOOLS

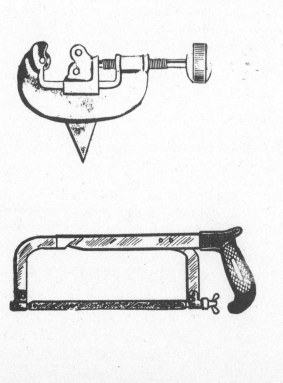

EAKY FAUCETS

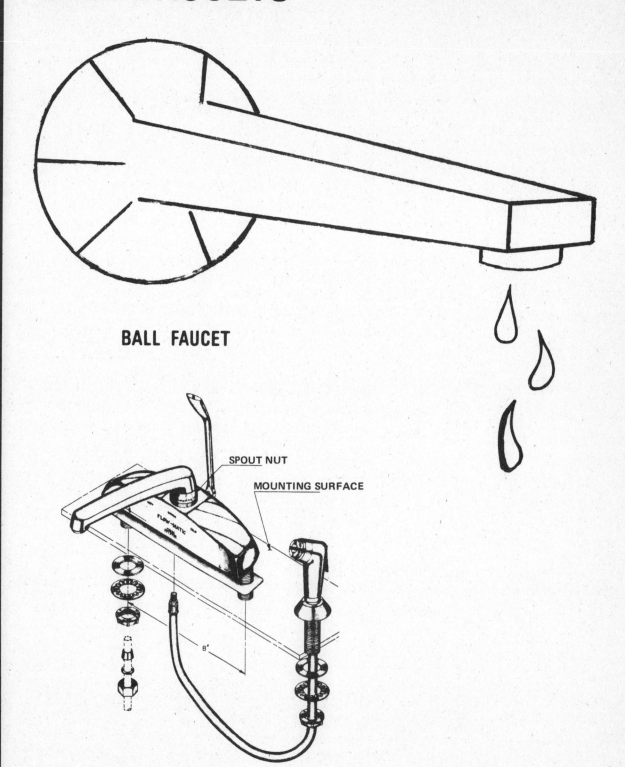

BALL FAUCET

SINGLE HANDLED BALL FAUCET

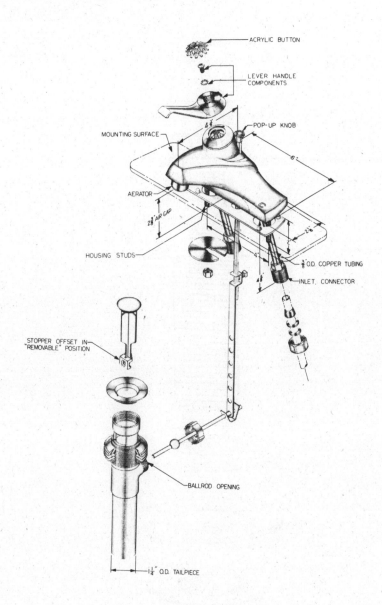

GLOBE FAUCET

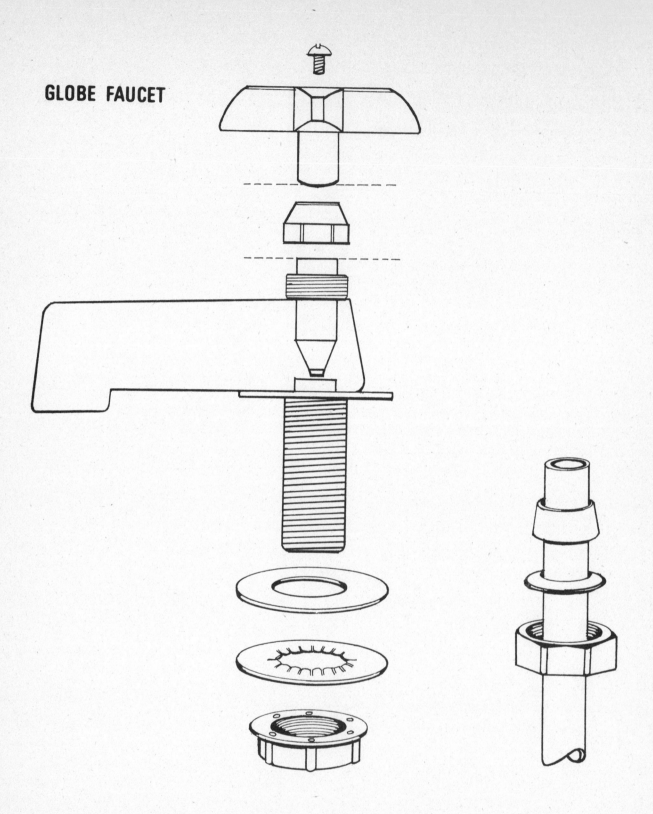

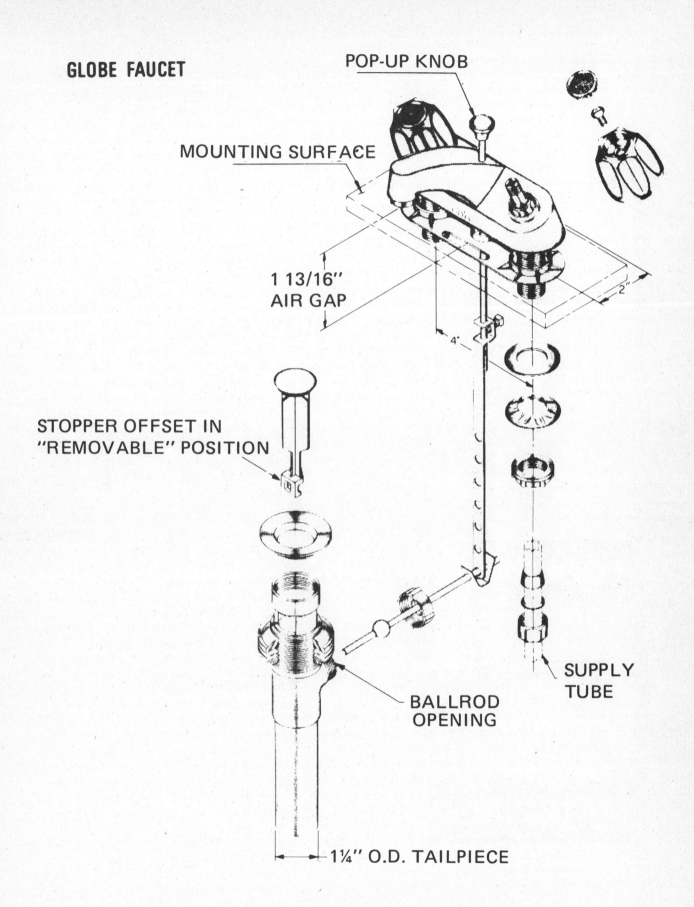

LEAKY FAUCETS

GLOBE FAUCET: Leak from under handle

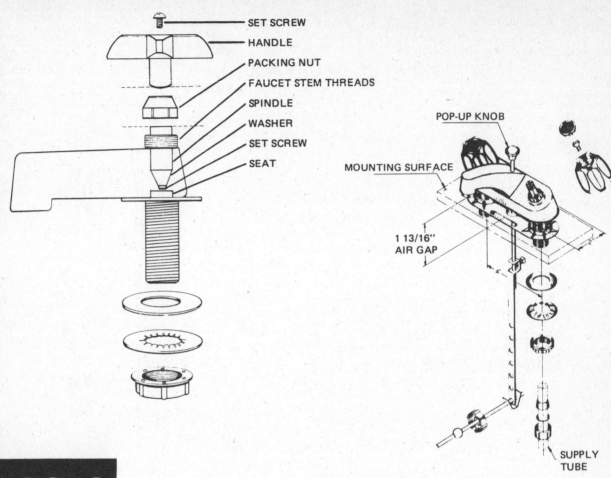

TOOLS

MONKEY WRENCH

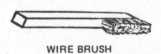
WIRE BRUSH

MATERIALS

Graphite packing
Adhesive tape

INSTRUCTIONS

The common globe faucet is the basic shutoff valve. It is found on most basement and outdoor water outlets, and on older sink types. It may leak from the mouth of the faucet, or from under the handle.

1 · If you wish to protect the finish on a globe faucet, wrap adhesive tape around the packing nut.

2 · Try to tighten the packing nut with a monkey wrench. Do not tighten it so tight that the handle won't turn freely. If this doesn't stop the leak, go to the next steps.

3 · Turn off the water to your faucet or valve. There is always a shutoff valve if you follow the water pipe back from the valve or faucet you are working on.

4 · Remove the handle to the faucet. The handle is usually held on by a set screw, which may be covered by a small cap, and can be removed with a screwdriver.

5 · Unscrew the packing nut with your monkey wrench.

6 · Scrape out the old packing from the inside threads of the packing nut, being very careful not to nick the threads.

7 · Use a wire brush to remove the packing from the faucet stem threads.

8 · Check both the threads inside the packing nut and the faucet stem threads, for stripped threads or nicks. If either one is in bad shape it should be replaced.

9 · To repack your packing nut, use graphite packing around faucet stem. In an emergency, packing can be improvised from plumbers lamp wicking, by wrapping it along the stem.

10 · Retighten the packing nut.

11 · Replace the handle of the faucet. If you have dual handles on a mixing faucet, make sure that you have the hot water handle on the hot water tap, and vice versa.

12 · Turn the water back on.

LEAKY FAUCETS

GLOBE FAUCET: Leak from end of spout

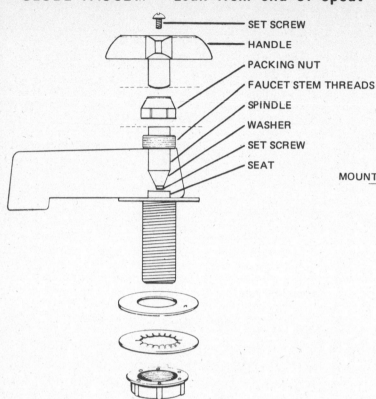

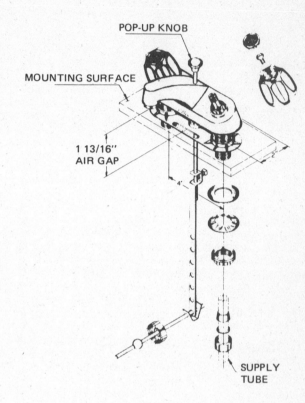

TOOLS

MONKEY WRENCH

SCREWDRIVER

MATERIALS

Possibly new seat
New washer - *may be flat or cone shaped. Be sure it is for hot and cold water.*

INSTRUCTIONS

1 · Turn off the water to your faucet. There will be a shutoff valve somewhere along the waterline leading to the faucet you are fixing.

2 · If you wish to protect the finish on your faucet, wrap adhesive tape around the packing nut.

3 · Use your monkey wrench to unscrew the packing nut.

4 · Remove the spindle by lifting it out while turning it in the direction you would turn the handle to turn the water on.

5 · The washer is held to the bottom of the spindle with a set screw. Take out the screw and remove the washer.

6 · Scrape off any pieces of the washer that might stick to the spindle, but be careful not to nick the spindle surface.

7 · Put a new washer on. Make sure that you use the same size and shape of washer as the one you took off.

INSTRUCTIONS

8 · Check the inside of the faucet body with a flashlight to make sure that none of the washer is sticking to the seat. Check the seat for nicks or corrosion. If the seat is in bad shape, it may have to be replaced. Many faucets have a removable seat. When you look into the body of the faucet you will see a hexagonal or square hole, if the seat is removable. A screwdriver, or an Allen wrench, of the correct size, will allow you to screw out the seat. Bring it to the store for an exact replacement. If you don't have a replaceable seat, you can either try to grind down the old seat with a seat grinder, or replace the faucet. In most cases it would be easier just to buy a new faucet, with a removable seat. The expense of a seat grinder is not really worth the price because it is seldom truly effective for long.

9 · After doing what you had to do with the washer and seat, replace the spindle assembly by turning it into the body of the faucet.

10 · Retighten the packing nut. Check the packing for leakage after you have it in, and repack it if necessary.

11 · Replace the handle.

12 · Turn the water back on.

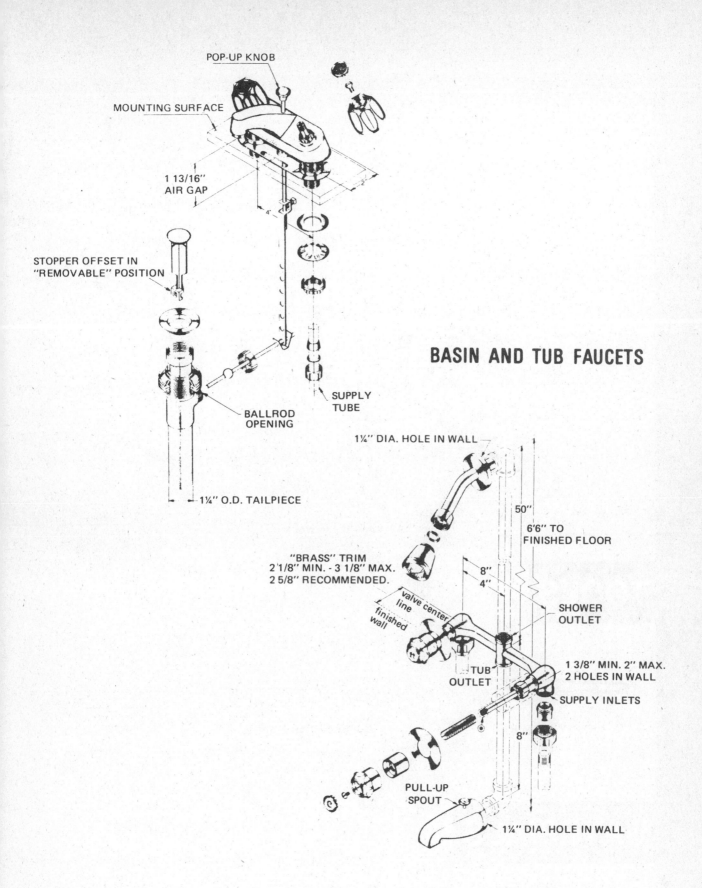

LEAKY FAUCETS

BASIN OR TUB FAUCETS: Leak from under handle - replacing packing

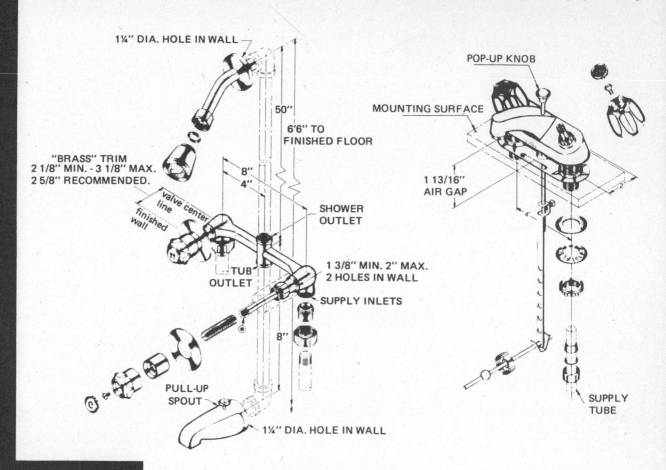

TOOLS

MONKEY WRENCH

SCREWDRIVER

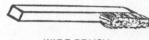
WIRE BRUSH

MATERIALS

Graphite packing
Adhesive tape

INSTRUCTIONS

Most of the newer basin and tub faucets shouldn't give you any trouble with leaks around the handle. Many don't have a packing nut to worry about on the spindle shaft. All faucets which work with a neoprene diaphragm, instead of with washer, can have a leaky handle fixed just by replacing the diaphragm. If you have to replace packing nut, however, it is easily done.

1 · Remove the handle of the faucet. It will probably be held on by a set screw, which may be covered by a cap.

2 · Remove the escutcheons, if there are any. These may also be held on by a set screw, or they may screw or pry off.

3 · If there was no escutcheon, you may want to wrap adhesive tape around the nut to protect the finish.

4 · The packing nut is the one furthest from the wall. Use a smooth jawed monkey wrench to try to tighten the packing nut. Do not tighten the nut so much that the water will not turn on with the handle.

5 · If tightening the packing nut did not spot the leak turn off the water to the faucet from the shut-off valve further along the water supply line.

6 · Use your monkey wrench to loosen and remove the packing nut.

7 · Use a wire brush to remove the old packing from the stem of the spindle, and wipe out the old packing from the inside threads of the packing nut. Check the threads in both places for nicks or stripped threads.

8 · If the threads are in good shape, repack the stem with string type graphite packing, and push packing nut down over packing. If the threads are badly stripped, you may have to replace the piece that is bad.

9 · Retighten the packing nut.

10 · Turn the water back on, to test the packing for leakage.

11 · Replace the escutcheon, then the handle.

LEAKY FAUCETS

BASIN AND TUB FAUCETS: Leak from end of spout

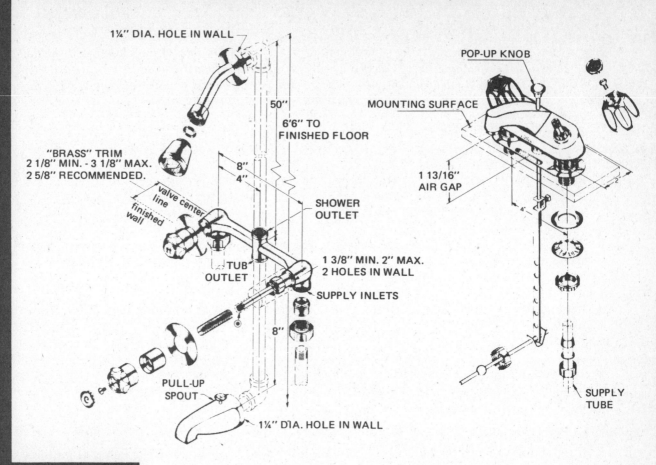

TOOLS

SCREWDRIVER

ALLEN WRENCH

MONKEY WRENCH

MATERIALS

New washer *Hot & Cold Water*
Possibly a new seat
Adhesive tape
The washer should be for both hot and cold water

INSTRUCTIONS

Many newer models of faucets on basins and tubs are different from the globe faucet. They require a little figuring out before you can fix them. It is a good idea to keep on hand the manufacturer's schematic diagram of your fixture. They come with every fixture, and if you don't have one they can be secured by writing or visiting the manufacturer or your local distributor.

1 · Turn off the water to the faucet.

2 · Take off the handle of the faucet. It is usually held by a set screw, which may be hidden by a cap.

3 · If there is an escutcheon covering the nuts, take it off. It may be held by a set screw, or it may twist or pry off.

4 · Look at the nuts which hold the spindle shaft to the sink or tub. Unless the works have been plastered and tiled over, the nut closest to the wall or sink will be a lock nut, which holds the fixture to the sink. There is another lock nut on the other side of the wall or basin surface. The next nut out from the surface will be the body nut. If there is a third nut it will be the packing nut.

INSTRUCTIONS

5 · If there is a body nut and a packing nut, use your monkey wrench to loosen the body nut until you can remove the spindle shaft. If there is just one nut, loosen it until you can remove the spindle shaft.

6 · Remove the washer at the end of the spindle. If there are two washers, with one at the end and another half way up the shaft, remove them both. Take the washers to the store with you to insure that you get the proper replacement parts. There may be a neoprene diaphragm instead of a washer. Replace it as you would a washer.

7 · While you have the body of your fixture open, look inside with a flashlight and check the condition of the seat. If the seat has parts of the old washer stuck to it, try to clean it off. If the seat is nicked or corroded you may have to replace it. The hole in the center of the seat is either hexagonal or square. If it is a hex, it can be removed with the proper sized Allen wrench, although a screwdriver may work. The square ones can be taken out with a screwdriver.

8 · Reinsert the shaft of the spindle, and tighten the body nut.

9 · Turn the water back on to check for leakage from the spindle, and to make sure that the faucet mouth has stopped leaking.

10 · Put the escutcheon back on, if there was one.

11 · Put the handle back on. Make sure that you have the hot water handle on the hot water valve, and that the handle is properly aligned with the other handle.

SINGLE HANDLED BALL FAUCETS

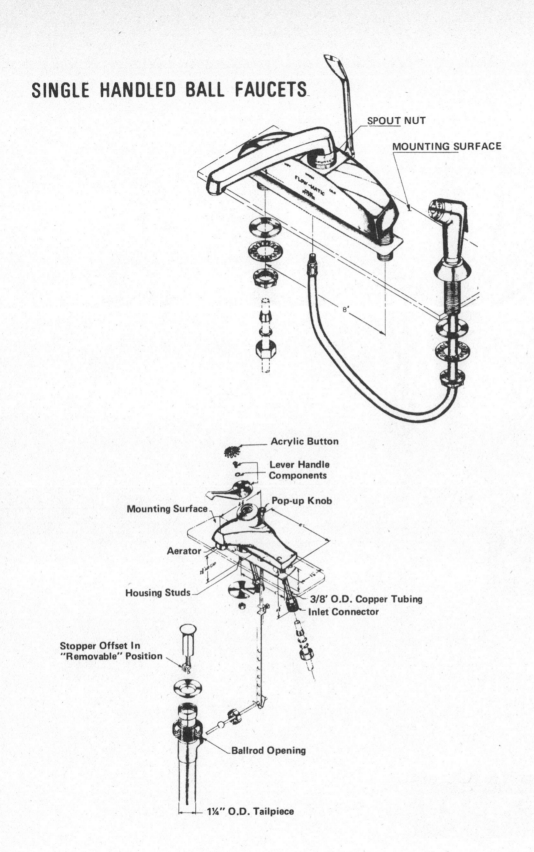

LEAKY FAUCETS

SINGLE HANDLED BALL FAUCETS: Leak from under the handle

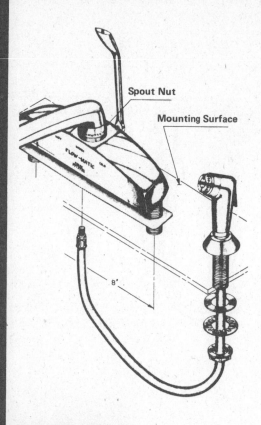
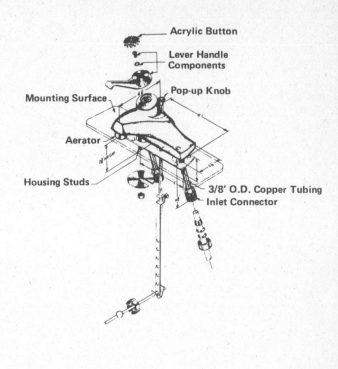

TOOLS

SCREWDRIVER

STRAP WRENCH

MATERIALS

None

INSTRUCTIONS

These faucets are of many different types, and you should take steps to secure a schematic diagram from the manufacturer of whatever type of model you own. These diagrams come with the faucet assemblies when they are bought and should be kept whenever you buy a new assembly. If you cannot find a schematic diagram, follow the steps outlined below, but be prepared for slight differences from what is described here.

1 · Turn off the water to the faucet.

2 · Take off handle. The handle will be held on by a set screw. You may need a very small screwdriver to loosen this screw.

3 · If your faucet leaks from under the handle, you should be able to stop it by tightening the adjusting ring with strap wrench.

4 · Put handle back on.

5 · Turn on the water.

LEAKY FAUCETS

SINGLE HANDLED BALL FAUCET: Leak from end of spout

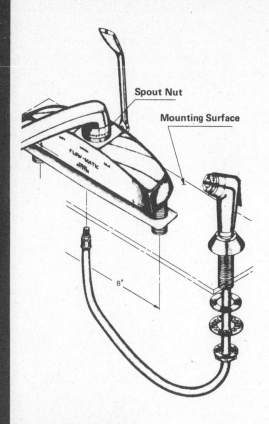
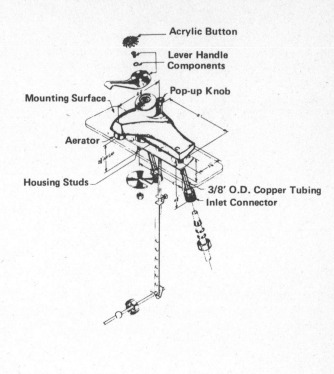

TOOLS

SCREWDRIVER

MONKEY WRENCH

MATERIALS

Adhesive tape *to protect finish*
New rubber seats and springs
Possibly a new ball
  Disassemble and take these parts with you when you go to buy new ones

INSTRUCTIONS

1. Turn off the water to the faucet.

2. Take off the handle. The handle will be held on by a set screw. You may need a very small screwdriver to loosen this screw.

3. Put tape around cap assembly and use your monkey wrench or strap wrench to unscrew and remove the assembly.

4. There will be a stem sticking up from the base. This is the ball stem. Pull up on this stem to remove the cam and ball assembly.

5. Lift out the rubber seat and springs from the pockets on the inside of the body of the faucet.

6. Replace these two rubber seats and springs.

7. Check the ball, and if there is any evidence of wear or roughness around either of the two small holes in the ball, it will have to be replaced also.

8. When reassembling, look for a pin on the inside of the hole in the body of the faucet. Make sure that the slot on the side of the ball is inserted over this pin. There may also be a slot on the side of the body. Make sure that the lug on the side of the cam is inserted into this slot. This is one area that may differ from model to model.

9. Replace cap assembly. When you have the assembly back on, tighten the adjusting ring on the top of the cap.

10. Replace the handle.

11. Turn the water back on.

LEAKY FAUCETS

BALL FAUCET: Leak from the top or bottom of the spout

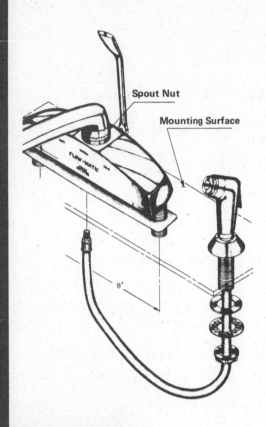
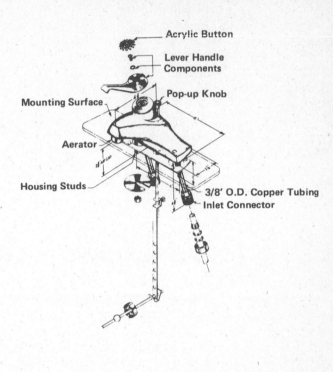

TOOLS

SCREWDRIVER

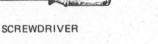
STRAP WRENCH

MONKEY WRENCH

MATERIALS

New body "O" rings
Adhesive tape, to protect the faucets finish

INSTRUCTIONS

1 · Turn off the water to your faucet.

2 · Take off the faucet handle. It should be held on by a set screw. You may need a very small screwdriver to remove this screw.

3 · Put tape around bottom of cap assembly and use your monkey wrench or strap wrench to unscrew and remove cap assembly.

4 · Pull up on the ball stem and remove the cam and ball assembly.

5 · Take the rubber seats and springs out of the pockets on the inside of the body of the spout.

6 · Slowly turn and pull up on the spout to remove it.

7 · Take off the two body "O" rings and replace them.

8 · Replace the spout by pushing it back onto body, while slowly turning it.

9 · Put the two rubber seats and springs back into the holes on the inside of the body.

10 · Put ball assembly back into body, making sure that the slot in the ball is inserted over the pin on the inside of the body.

11 · Put cam back on top of the ball stem, making sure that the lug on the side of the cam is inserted into the slot on the side of the body.

12 · Replace cap assembly. Make sure to tighten the adjusting ring to keep water from leaking under the handle.

13 · Replace handle.

14 · Turn the water back on.

LEAKY FAUCETS

BALL FAUCET: Cleaning clogged diverter assembly

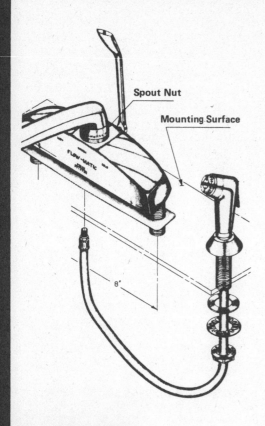
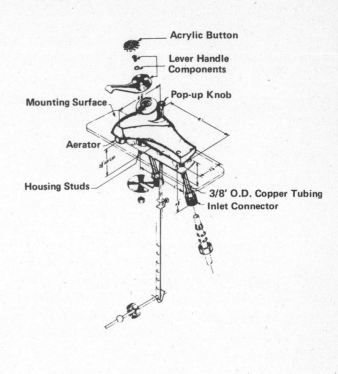

TOOLS

SCREWDRIVER

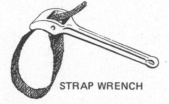
STRAP WRENCH

MONKEY WRENCH

MATERIALS

Adhesive Tape: to protect the faucets finish

INSTRUCTIONS

If the ball faucet seems to be getting clogged — that is, if the water doesn't run at all, or with less volume of water than usual, you may have to clean the diverter assembly. These models of faucets are more easily clogged by water sediments than any other types of faucets.

1 · Follow the same disassembly procedure as in repairing leaks from top or bottom of spout body, up until Step 6.

2 · After removing the spout, pull out the diverter from the hole in the outside of the body.

3 · Thoroughly clean the diverter with water.

4 · Replace the spout by pushing it back onto body, while slowly turning it.

5 · Put the two rubber seats and springs back into the holes on the inside of the body.

6 · Put ball assembly back into body, making sure the slot in the ball is inserted over the pin on the inside of the body.

7 · Put cam back on top of the ball stem, making sure the lug on the side of the cam is inserted into the slot on the side of the body.

8 · Replace cap assembly. Make sure to tighten the adjusting ring to keep water from leaking under the handle.

9 · Replace handle.

10 · Turn the water back on.

LEAKY FAUCETS

REPLACING BASIN or KITCHEN DECK FAUCET

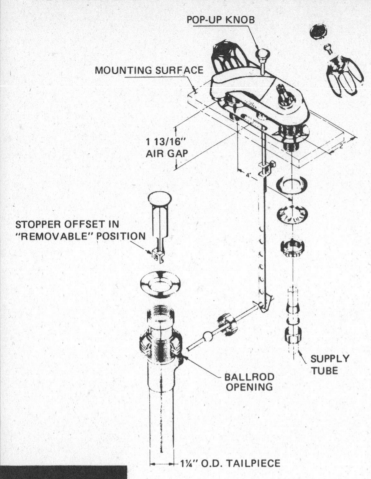
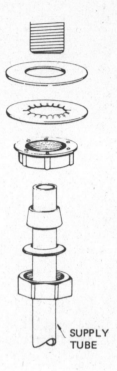

TOOLS

Basin wrench can be rented -
(It is vital for reaching up behind basin or sink)

MATERIALS

New faucet(s)

INSTRUCTIONS

1 · Turn water off at nearest shut-off valve.

2 · Disconnect first nut under basin or sink, using basin wrench to reach it and turning counter clockwise.

3 · Disconnect second nut, located higher up, with basin wrench.

4 · Lift out faucet. (Do the procedure again on second faucet if you have more than one).

5 · Put the new faucet or faucets in place.

6 · Do steps 2 and 3 in reverse, tightening higher nut first, then lower one, turning both clockwise.

7 · Turn water back on.

TOILETS- MAINTENANCE

The toilet, often referred to as a commode or water closet, is one of the commonest causes of plumbing woes. It is also one which the economy-minded homeowner should be able to fix himself. Everything from maintenance to installation can be handled fairly simply.

A toilet needs no care beyond occasional cleaning. Check occasionally for leaks. Open the top of the tank to check the following:

1 If the flush ball or flapper is coated with slime or mineral deposits picked up from the water supply, clean it thoroughly to insure a good seat over the flush valve hole. If there are mineral deposits, it may be necessary to replace the ball or flapper.

2 Check the flush valve seat for slime or mineral deposits. Keep it as clean as possible.

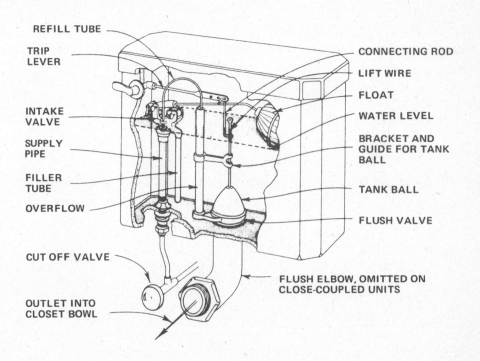

3 Check for bent lift wires or crushed, corroded flush ball or flapper. Replace if necessary.

4 Listen for sounds that continue too long after the toilet is flushed. They indicate water is being wasted.

5 Occasionally touch a piece of paper to the dry, inside, back of the bowl above the water line. If the paper gets wet it may indicate a slow leak that can cost money in water bills.

6 Check the refill tube to be sure it is not clogged by slime or mineral deposits.

7 Check that the water level is about ¾ inch below the top of the overflow tube.

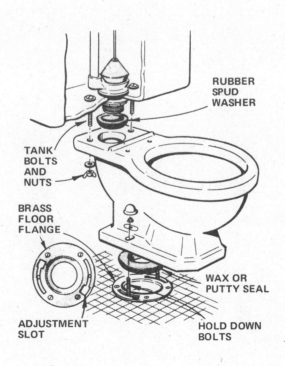

TOILET TANK LEAKS

REPLACING FLUSH BALL

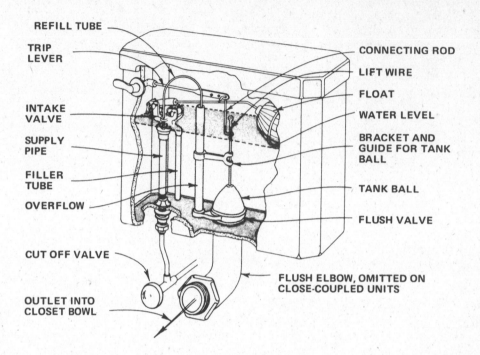

TOOLS

SANDPAPER

SOAP PADS

STEELWOOL

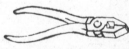
SIDE-CUTTING PLIERS

MATERIALS

New flush ball of proper size
New life wires (brass are best)

INSTRUCTIONS

If the toilet constantly sounds as though it has just been flushed although it has not been in use, or if you find a silent leak with your paper test (step 5 in maintenance) follow the repair sequence that follows.

Check to see if the flush ball or flapper needs cleaning or replacing - or if the seat of the flush valve hole needs cleaning. (A flapper valve will seldom if ever need to be replaced).

1 · Turn off the water to the toilet with the stop valve under the tank or in the basement.

2 · Flush the toilet to remove water from tank.

3 · Unscrew the flush ball from the lower lift wire. If it does not unscrew, use wire cutters to snip off and remove the lift wire.

4 · Scour the seat for the flush valve with a soap pad, steel wool or fine sandpaper. Rinse off seat.

5 · Attach a new ball. If you change lift wires at the same time, attach wires first, then the ball.

6 · Check that the flush ball is centered directly above the flush valve seat and that it drops smoothly down to the seat. If it doesn't, adjust the guide arm so that it will.

7 · To insure a good seal between the flush ball and the seat, smear a little vaseline or other lubricating jelly around the rim of the seal where it meets the flush ball.

TOILET TANK LEAKS

REPAIRING FLOAT APPARATUS

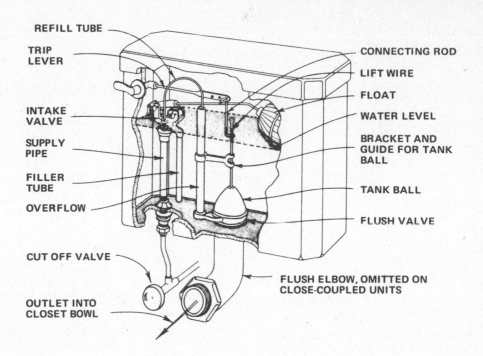

TOOLS

None

MATERIALS

New float (possibly) (Follow first steps of instructions before buying)

INSTRUCTIONS

If your tank is leaking and the flush ball apparatus seems all right, lift up the float arm. If the noise stops or the leakage stops, follow instructions below.

1 · Check to see if the float arm or the float is getting caught on anything in the tank.

2 · If the float arm and float are not getting caught, shut off water to the tank with the stop valve underneath the tank or in the basement.

3 · Unscrew the float from the float arm and shake it to see if it has any water in it. If there is water in it buy a new float and screw it onto the float arm.

4 · If raising the float arm stops the leakage, yet the preceding steps have not cured the problem, your water level might be to blame. Try bending the float arm downward from the middle. Hold the middle and push the float end down until the end bends and the float rides about a half inch lower in the water.

5 · Turn water back on and see if the problem has been solved.

TOILET TANK LEAKS

REPAIRING FLOAT VALVE

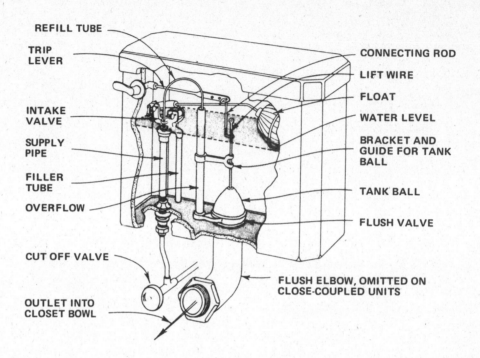

TOOLS

SCREWDRIVER

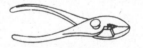

PLIERS

MATERIALS

Two washers - (Bring old washers to supply store to match for correct size)

INSTRUCTIONS

If the first sections in this series on toilet tanks have been tried and have not worked, the problem may be a balky float valve. This can be fixed by replacing the washers.

1 · Turn off water to the tank with the stop valve underneath the tank or in the basement.

2 · Flush the toilet to drain the tank.

3 · Unfasten the two pivot screws which hold the end of the float arm to the float valve.

4 · Remove the float arm and its linkage by sliding it out of the slot in the plunger top.

5 · Lift out the plunger.

6 · There is a rubber or leather washer screwed onto the bottom of the plunger, and another one about half way up on the plunger. Remove these with a pair of pliers and replace them.

7 · Replace the plunger.

8 · Slide the linkage end of the float arm back through the slot of the plunger and re-fasten the two screws which hold it in place.

9 · Be sure the refill tube is feeding into the overflow pipe.

10 · Turn the water back on.

FLUSHING PROBLEM

If you have trouble with the trip handle on the toilet when it is flushed (such as having to hold the handle down while it flushes), the problem is with the lift wires. The lift wires are not raising the ball high enough to keep it from being sucked back down by the rushing water. Use a pair of pliers to straighten the upper life wire, then rebend it so that it is shorter.

The trip handle never needs work beyond an occasional tightening of the nut on the inside of the tank behind the handle. Use an adjustable open-end wrench and turn the nut counter-clockwise.

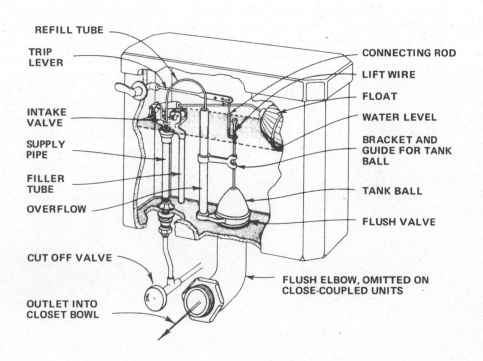

TOILET BOWL

LEAK BETWEEN TANK AND BOWL:

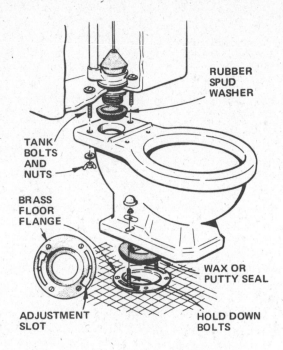

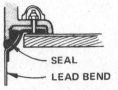

TOOLS

SCREWDRIVER

MONKEY WRENCH

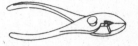
PLIERS

MATERIALS

New rubber spud washer
Possibly
New tank (possibly)
Epoxy glue (possibly)
Petroleum jelly

INSTRUCTIONS

Many new toilets are one piece units. Older facilities have separate bowl and tank units. These are connected and sealed by a rubber spud washer. This washer should seal the connection from leaks indefinitely, but if the tank was improperly installed, or if the tank has been empty or unused for a long time, it may become dry and crack, or corrode. You may want to replace just the tank if it has been cracked.

1 · Shut off the water to the toilet, at shut off valve below the tank or in the basement.

2 · Flush the toilet to drain the tank.

3 · Disconnect the pipes bringing water to the water supply intake pipe in your tank. There is usually a take-up nut which can be disconnected with a wrench or pliers.

4 · Place some newspapers or old towels on the floor out of your way where you can put the tank when it has been removed.

5 · Disconnect the tank from the bowl. It is usually held on by two bolts and wing-nuts with washers. Be careful with the tank, unless it is already cracked or broken. It will be heavy and must be lifted straight up off the bowl. Place it on the newspapers or towel padding.

6 · After removing tank, inspect the rubber spud washer. If it is deteriorating or cracked, it will have to be replaced. If in good condition, check the porcelain of the tank and bowl for cracks. These cracks can often be repaired with an epoxy glue.

7 · Install a new spud washer if necessary, or check the seating on the old one. Put a little petroleum jelly around the rim of the spud once it has been carefully reset.

8 · Reset the tank on the bowl and retighten the wing nuts onto bolts. Do not forget the washers.

9 · Re-connect the water lines.

10 · Turn on water to refill tank.

TOILET BOWL

LEAK BETWEEN BOWL AND FLOOR:

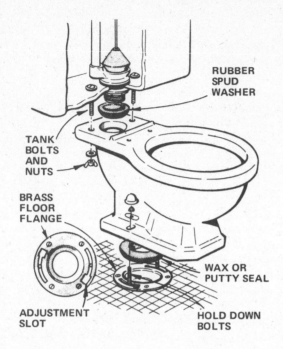

TOOLS

SCREWDRIVER

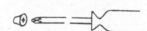
PHILLIPS SCREWDRIVER

PLIERS

ADJUSTABLE PIPE WRENCH

PUTTY KNIFE

MATERIALS

New wax bowl gasket
Petroleum jelly
Floor tiles and
Floor tile cement *(possibly)*

INSTRUCTIONS

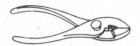

Some of the newer toilets may connect to the wall, rather than to the floor, but in either case the same basic procedure may follow.

1 · Shut off water to the tank by turning shut-off valve right below tank or in the basement.

2 · Flush the toilet to drain the tank.

3 · To avoid a mess, siphon or mop up as much of the remaining water in the bowl as possible. Place newspaper or old towels as padding for the tank and bowl to rest on after you have removed them, being sure they are well out of your way.

4 · Disconnect the water lines from the tank. There is usually a take-up nut which can be removed with a wrench or pliers.

5 · Unless you have a one piece unit, disconnect the tank from the bowl. The tank is usually held on by just two bolts. Place the tank on the padding, out of the way.

6 · Remove the seat and cover from the bowl. You may need a Phillips head screw driver for this.

7 · Remove the porcelain or china bolt caps, if any, and remove the bolts holding the bowl to the floor flange. Rock or jar the bowl enough to break the seal at the floor connection. Lift the bowl straight up to remove it from the bolts. Then place it upside down on pad or papers or towels.

8 · Use a putty knife or other scraping tool to remove the old wax or rubber bowl gasket. It may have been sealed with a putty compound. Remove this material from both the bottom of the bowl and from the floor flange.

9 · Place a new toilet bowl gasket around the bowl horn on the bottom of the bowl. The now standard wax gasket is the best.

10 · Remove any scraping or plaster from the area which will be covered by the toilet. Taking up the bowl may tear up tiles from the floor which will have to be cemented after you have seated the bowl.

INSTRUCTIONS

11 · Turn the bowl right side up again and lower it slowly back onto the floor flange. Try to do this as carefully and as evenly as possible. You may need help. When it is on the flange, press straight down on the bowl with your full weight and twist it slightly from side to side to be sure of a correct seating. It should be perfectly even. You may even want to check it with a level.

12 · Re-install the tank. It is a good idea to put some petroleum jelly around the rim of the rubber spud washer to insure a good seal between the tank and bowl. Do not forget the washers when putting on the wing nuts. Tighten the nuts hand tight, then using pliers to turn them another time.

13 · Reconnect the water line to the tank.

14 · Turn the water on and flush the toilet once to test the seal for leaks.

15 · Re-attach the seat and cover any bolt caps that may have been there.

16 · Repair any floor damage with plaster of paris, or replace tiles.

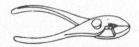 

HEATING SYSTEMS

A heating system may work by hot water, steam or hot air, and it may be run by electricity, gas, oil or coal. Each type of system has its good and bad points. The primary consideration for the homeowner must be the availibilty of the fuel service. The companies which service and supply energy needs vary greatly in costs and service from one area to another. If you are thinking of changing your heating system, these costs should be carefully reckoned along with the actual installation costs. Whatever system your house has, the first priority is to make it as efficient and effective as possible. This means regular cleaning of the system.

HOT WATER SYSTEMS

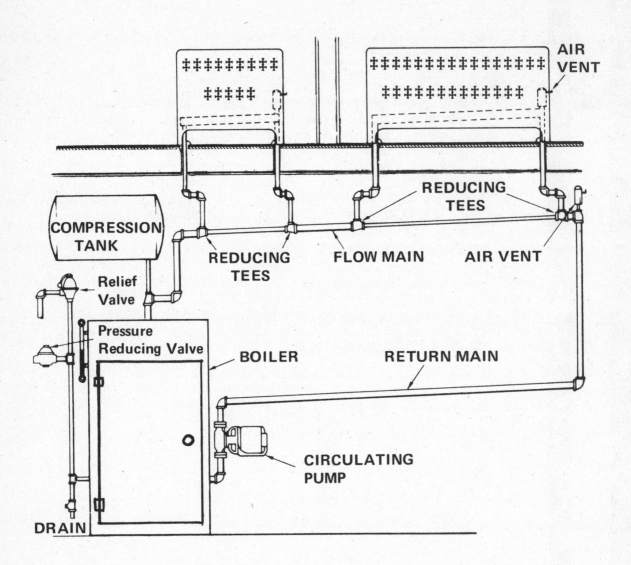

STEAM HEATING SYSTEMS

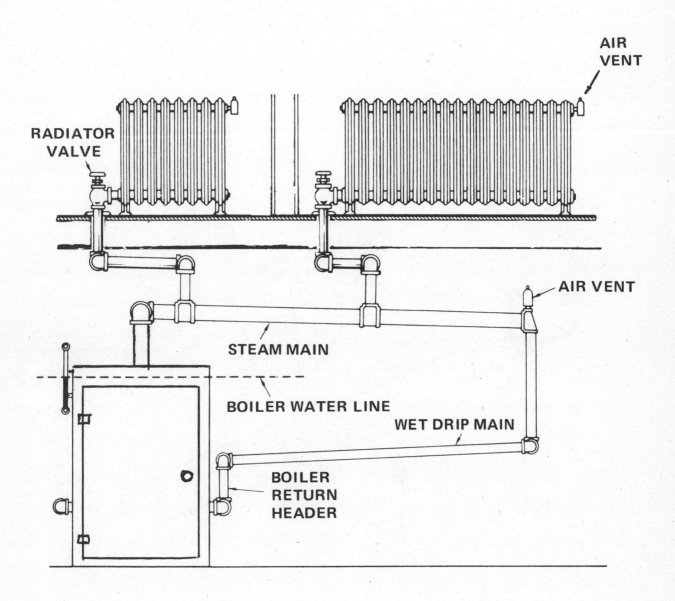

Sweating Pipes

If you have sweating pipes, caused by the condensation of moisture from the air on exposed cold water pipes, the condition can be alleviated by wrapping the pipes with insulation.

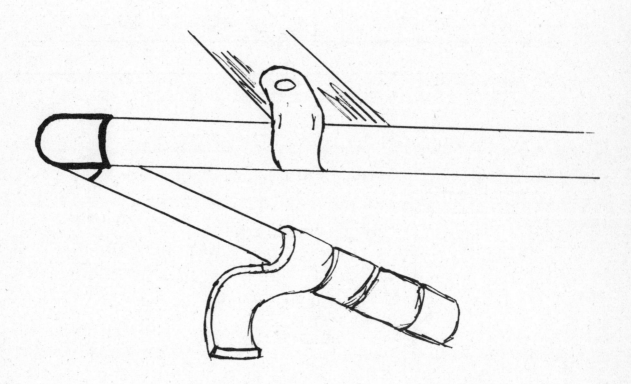

INSTRUCTIONS

Water Hammer

When a water outlet has been running full blast, and then the water is shut off all at once, you may become aware of loud hammering noises in the pipes. Your water supply system runs on water pressure, usually 60-70 p.s.i. (pounds per square inch). When water is shut off suddenly, the velocity of the incoming water causes the pressure to mount, to as much as 100 p.s.i. Looking for a way out, it slams back and forth along the length of pipe, making a racket.

Metal pipes can withstand this pressure, although some plastics cannot, but the noise is annoying. For all plastic pipe systems, and noisy metal pipe systems, add on 12 or 18 inch air chambers near the shut-off valve. These are empty pipes, capped off on top, which give the pressure something to slam up against without making noises and avoiding the build-up of pressure. The air chambers should be placed as close as possible to each fixture's water supply outlet.

Prevent A Leak

Always turn off the water to the washing machine when not in use. If you don't, the pressure on the hoses will eventually cause hoses to give way with a resultant flood.

HEATING SYSTEMS

YEARLY CLEAN-OUT with STEAM HEAT

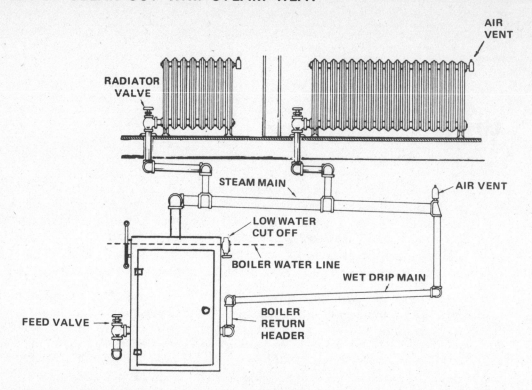

GOING ON VACATION? WHENEVER THE VACATION IS WARM WEATHER, TURN OFF OIL BURNER WHILE AWAY. OTHERWISE, IT RUNS, MAKING HOT WATER UNNECESSARILY.

TOOLS

SCREWDRIVER

HOSE

BUCKET

MATERIALS

None

INSTRUCTIONS

A heating system may work by hot water, steam or hot air, and it may be run by electricity, gas, oil or coal. Each type of system has its good and bad points. The primary consideration for the homeowner must be the availibilty of the fuel service. The companies which service and supply energy needs vary greatly in costs and service from one area to another. If you are thinking of changing your heating system, these costs should be carefully reckoned along with the actual installation costs. Whatever system your house has, the first priority is to make it as efficient and effective as possible. This means regular cleaning of the system.

1 · Shut off the emergency switch on the furnace.

2 · Shut off cold water feed valve to boiler. This valve should be off anyway.

3 · Run off water from drain valve at lowest point near the furnace and boiler. Either drain the water into a bucket, or attach a hose and drain it with that. Run until the water completely stops flowing. Then shut off valve.

4 · Open the cold water feed valve and fill the boiler about half way.

5 · Shut off the cold water feed valve.

6 · Drain boiler again. This second draining will clean out the sludge that may have been left on the inside of the boiler.

7 · Shut off drain valve.

8 · Reopen cold water feed valve. Fill until the water glass gauge on the furnace has water up to the line on its face. Then shut off cold water feed valve.

9 · Turn furnace emergency switch on.

HEATING SYSTEMS

BANGING RADIATORS - STEAM HEAT

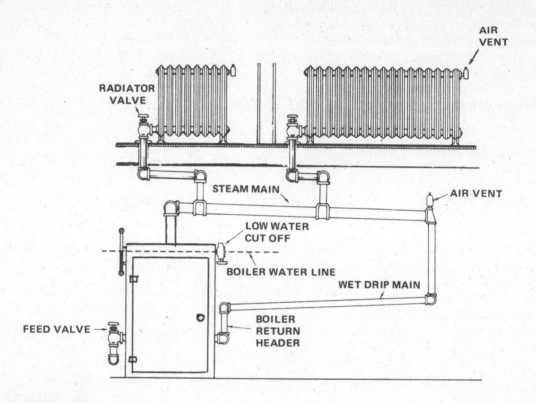

TOOLS

SCREWDRIVER

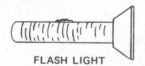
FLASH LIGHT

ALLEN WRENCH

MONKEY WRENCH

MATERIALS

New Washer
New valve seat (possibly)
Graphite packing

INSTRUCTIONS

Sometimes, especially in older steam heat systems, the radiators may start banging. This is caused by water going into the condenser coils from the cold water feed valve. It means you either have a leaky stop cock on the feed valve or a leak in the condenser coils.

1 · Check the water glass gauge on the furnace. This gauge is usually about 6 inches long and you should have a water line at about 2½ inches. If you cannot see a water line, assume that the glass gauge is filled to overflowing. If the water line is steady at 2½ inches, the problem may be a leaky condenser coil. Fix the stop cock anyway, then if the banging persists, call a repairman.

2 · Try to tighten the cold water feed valve. If it wasn't closed all the way, the problem may be solved that simply.

3 · Turn off the emergency switch on the furnace.

4 · Find the shut off valve on the water pipe which leads to the cold water feed valve and turn it off.

5 · Take off the handle to the cold water feed valve. It is usually held on with a set screw.

6 · With the monkey wrench, unscrew the packing nut on the valve.

7 · Lift out spindle.

INSTRUCTIONS

8 · On the bottom of the spindle there is a washer, held on by another set screw. Take off the washer and replace with new washer.

9 · If possible, look inside of the body the valve with a flashlight to check the seat. If the washer looks in good shape, the seat may need replacing. Many of the newer valves have removable seats, which can be taken out with a screwdriver or an Allen wrench. Replace it. (Many plumbers would recommend grinding down the seat with a seat-grinder, but it usually easier to replace the whole valve).

10 · When you have replaced the washer, and perhaps the seat too, put the spindle back into the valve.

11 · Retighten the packing nut. It will probably need new packing. Use graphite packing.

12 · Replace the handle.

13 · Make sure the valve is shut tight.

14 · Turn on the water to feed valve from the valve "upstream" of the feed valve.

15 · Turn on the emergency switch on the furnace.

HEATING SYSTEMS

YEARLY CLEAN-OUT with HOT WATER HEAT

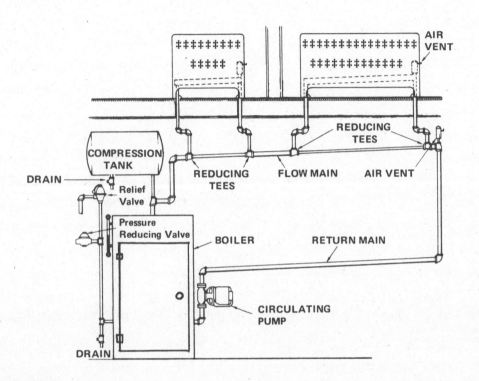

TOOLS

SCREWDRIVER

HOSE

BUCKET

MATERIALS

None

INSTRUCTIONS

HOSE

A hot water system should be cleaned out annually. The basement expansion tank should be drained periodically during the heating season. The regulator and gauges should be checked occasionally.

1 · Shut off the furnace emergency switch.

2 · Shut off the cold water feed valve.

3 · Attach hose to the drain valve at the lowest point of the furnace, or put bucket beneath if it is high enough off the ground.

4 · Let the water drain until it stops.

5 · Go to the radiator at the high point in your house, third floor, second floor or upper level in a split level, and open the air valve on each radiator on that level. This can be done with a screwdriver, a special key for that purpose, or handle, depending on the kind you have.

6 · Work your way down through the house, opening up all air valves on each of the radiators on each successive level of the house.

7 · When all the air valves have been opened, and all the water has stopped flowing from the drain faucet, shut off the drain faucet.

INSTRUCTIONS

8 · Go back to the top level of the house and work your way back down, closing all the air valves on all the radiators.

9 · Open up the cold water feed valve on the boiler and let it fill to the proper pressure gauge reading. When all radiators are filled, pressure will be 12-14 pounds.

10 · Go back up to the first level of radiators in the house. Go from one radiator to the next and open the air valve. Stay at that radiator until water starts coming out of the valve. Then turn off the air valve. When you have done this with each radiator at the first level of the house, go up to the next level and continue until all the radiators have been drained of air.

11 · Go back to the basement and check the pressure again. The feed valve should be left on in a hot water system. The regulator valve will take care of shutting off the feed water at 12-14 pounds.

12 · Turn the furnace emergency switch back on.

13 · The day after you have cleaned out the system, go through the house and check for air in the radiators again by opening up the air valves. This should be checked periodically throughout the heating season.

NOTE: If the hot water heating system has a faulty regulator, the pressure guage will read high. Over 30 pounds of pressure a cut-off valve breaks in, at which point the water will just keep draining into the floor drain. If this happens, call a plumber.

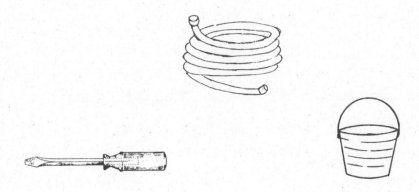

WATER SUPPLY SYSTEM

Your water supply system begins with a service entrance line, which takes the water from the water main to the water meter, unless you have your own water supply. From the meter the water is taken directly to your cold water outlets, and indirectly through your hot water tank, to your hot water outlets. The system works by pressure from the water main and is controlled by numerous shut-off valves.

Hot and Cold water lines usually run together, with pipes supplying whatever water is needed to the different fixtures. The toilet usually uses only cold water; a washing machine uses hot-cold water; basins and tubs use both.

Shut-off valves are an important part of the water supply system. When working on any part of the system, it is important to know where the shut-off valve nearest to your work can be located. Leaky valves can be repaired - as explained in the fixture segment of this book.

RECOGNIZING PIPE MATERIALS

You should always try to replace a pipe with a pipe made from the same type of material. If you do change the type of pipe, be sure you can use an adaptor coupling. In most cases, an adaptor will be necessary anyway because types of pipe are made in different diameter thicknesses.

YELLOW BRASS
When scratched, it will be a shiny gold color.

GALVANIZED STEEL
Heavier and thicker than any other pipe except lead, it is silver in color.

LEAD PIPE
Forget it. Lead pipe is too hard to handle. Even your plumber would rather replace them with something else.

FLEXIBLE POLYETHYLENE PLASTIC
Polyethylene will be found connecting fixtures to water supply pipes. It is for cold water only.

RECOGNIZING PIPE MATERIALS

PLASTIC
(Chlorinated Polyvinyl Chloride — CPVC).
Plastic is easily recognizable.

COPPER
When tarnished, copper pipe turns dark, but when you scratch the surface, it will be a shiny copper color.

RED BRASS
When scratched, red brass will resemble copper in coloring, but brass pipe is much thicker in diameter.

LEAKY PIPE BASICS

Most problems with leaky pipes are actually from leaks in connections between pipes. Most pipes will last a long time, depending on the chemical characteristics of your water supply. If one pipe is corroded to the point of leakage, it is likely the rest of the system is deteriorating.

Regardless of what type of pipe you have, do the following:
1 · Turn off the water supply to the connection.
2 · Drain the water from the pipe by opening the nearest outlet.
3 · Try to disconnect the coupling. Then realign the pipes and retighten the connection. In some cases this will be all that is needed.

In replacing a leaky pipe, the basic idea is to:
1 · Thread the two ends of the pipe you have created by cutting out the leaky section.
2 · Connect couplings to the threaded section with a length of pipe in between. This process varies with the materials which the pipes are made of.

Assuming that the entire system and the new section of pipe are of the same material, the process for each type follows.

LEAKY PIPES

Plastic or flexible polyethylene pipe Leak at connection

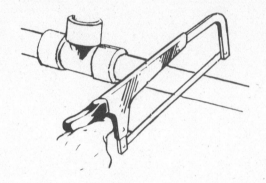

TOOLS

ADJUSTABLE HACK SAW

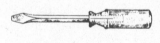
SCREWDRIVER

FILE

POCKET KNIFE

MATERIALS

A length of plastic pipe
Plastic pipe couplings
Sealing compound
Brush *one half the diameter of pipe with stiff non-synthetic bristles - Brush and compound may come in a kit.*

INSTRUCTIONS

Unless plastic pipe has burst from too high pressure or broken from outside, it should never leak except at joints.

1. Turn water off at shut-off valve near work area. Drain the pipe. Cut out the coupling that leaks, using a hacksaw. Try to make the cut flat across.
2. Use a small knife to clean the burrs from inside of the exposed ends of pipe.
3. The couplings for plastic pipes will necessitate cutting the replacement section slightly smaller than the cut-out section.
4. Use the brush to apply plastic solvent to the outside of the end of the pipe, to the depth of the coupling, and to the inside of the coupling. Immediately connect the fitting to the end of the pipe. Attach one fitting to one of the old sections of plastic pipe; the other fitting to one end of the replacement section of pipe.
5. Wait a minute or two to insure that the fittings are on firmly. Then make a dry run of putting the pipe section in place, without any solvent. When the plastic tubing is installed, it should have room for expansion. When hot water runs through plastic pipe, the pipe may expand in length. This gives you a little extra room to move the pipe when fitting in the new section - but make sure to always leave room.
6. Once you have checked the fit and taken the new section back off again, you should be prepared to work fast. Make sure all burrs are removed and the insides of the coupling are clean. Apply the solvent to the other end of the old pipe and to the inside of the fitting on the new section. Connect the two with a slight twist to make the seal firm. Then do the same with the two remaining sections.
7. The seal on plastic pipe couplings are stronger than the rest of the pipe, but they must be allowed time before they are tested. Wait two hours or more before testing, leaving water turned off until then.
8. After waiting several hours, turn the water back on, making sure the nearest outlet is open. Wait until the water fills the pipe and comes out of the outlet. Check the new fittings for leakage on and off for the first few hours. If no leak appears by then the system is okay.
9. One final word. We have mentioned that plastic pipe expands and room should be left for the expansion. There are special pipe hangers made for supporting plastic pipe which give enough room to allow for expansion. These hangers leave the pipe relatively loose. One should be placed every three feet or less.

LEAKY PIPES

PLASTIC PIPE: Changing to removeable coupling

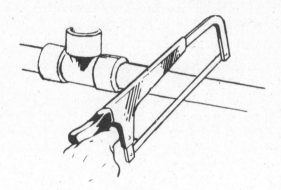

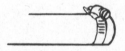 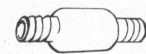 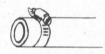

TOOLS

   OR

SCREWDRIVER   PLIERS

MATERIALS

Clamps or coupling *(Possibly)*
Plastic or polyethylene pipe

INSTRUCTIONS

There is a second type of coupling used for only cold water pipe fitting using plastic or polyethylene pipe. It uses threaded pipe and fittings with screw clamps, and needs no solvent. This type can be taken on and off a number of times with no replacement parts necessary. If a leak does occur, it may be fixed by tightening the clamp or the coupling. If this does not work, check which side of the coupling has the leak, and replace the pipe on that side.

1 · Turn off the water to the leak at the nearest shut-off valve. Let the water drain out of the nearest outlet.

2 · Unscrew the clamp to allow removal of the end of the pipe. If it doesn't pull or unscrew immediately, pour some hot water over the fitting and try again.

3 · Take off the hose and replace it. Tighten up the screw clamp. (Make sure you have the clamp on the hose before attaching the hose to the coupling). You may pressure test the fitting immediately by turning water back on.

Fixing a leaky hose on a dishwasher or washing machine would probably involve one of these types of connections.

LEAKY PIPES

METAL PIPES - Removing Soldered Pipe from Fitting

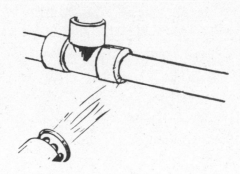

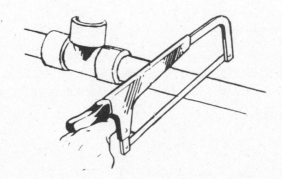

TOOLS

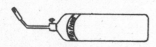
PROPANE TORCH

ADJUSTABLE HACK SAW

MATERIALS

Aluminum foil
Available water

74

INSTRUCTIONS

There are three types of fittings commonly used for metal pipes: soldered, flared and compression. Flare and compression fittings are very expensive but are easily used and can be taken apart and put together again with little difficutly. For the economy-minded homeowner, these types of fitting would probably be used only in places where soldering would be difficult.

1 · Turn off water to the fitting by turning off shut-off valve to the area. Drain the pipe of water at the nearest outlet.

2 · Cut the leaky pipe in the middle so you can remove one end at a time.

3 · Be sure that the area behind and around the fitting from which you want to remove the pipe is clear of flammable materials. Have water nearby in case of fire. It is a good idea to use heavy duty aluminum foil to shield the background.

INSTRUCTIONS

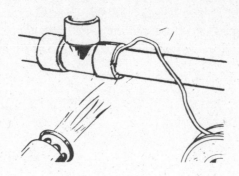

4 · Light the propane torch.

5 · Apply flame from torch to where the solder seals the fitting. Keep the flame moving back and forth. The pipe itself should never get so hot that you can't hold it 8 to 10 inches from where you are applying the flame.

6 · As you heat the solder, try to twist the pipe in the fitting. When you can turn the pipe in the fitting, twist and pull on the pipe to remove it.

WORKING WITH PIPE

CUTTING PIPE

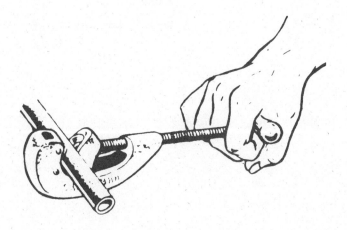

TOOLS

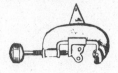
TUBE CUTTER

PIPE CUTTER

FILE

MATERIALS

Threading oil

INSTRUCTIONS

You may cut a pipe with a hacksaw but it may be worthwhile to buy or rent a pipe or tubing cutter. If you are working on pipe in a tight space, then you may have to cut with a hacksaw.

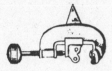  

1 · Mark the pipe where you wish to cut it.

2 · Using the pipe cutter, slip it onto the pipe where you marked it.

3 · Tighten the cutter down onto the pipe until it digs in.

4 · Turn the cutter around the pipe, one revolution and, applying threading oil one drop per revolution, or less.

5 · When you are nearly through the pipe it will probably begin to sag. Try to keep it as straight as possible until you cut all the way through.

WORKING WITH PIPE

THREADING PIPE

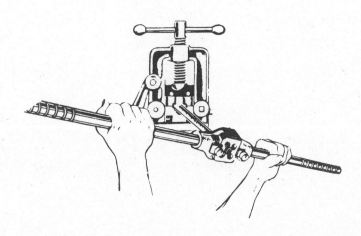

TOOLS

PIPE THREADER

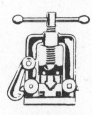
PIPE VISE

FILE

MATERIALS

Threading oil
Rag

INSTRUCTIONS

Note: Copper tubing seldom needs threading.

A pipe threader consists of a pipe die held in what is called a stock. You will have to be sure you have a die of the correct size for the pipe to be threaded.

1 · Place the pipe to be threaded into a vise.

2 · When you cut the pipe, you probably left a burr on the inside rim of the pipe. Use a reaming tool to remove burr.

3 · Set up the threader. Place it over the end of the pipe and turn the handle around the pipe clockwise. Bathe the pipe end with threading oil. When you first turn the threader, you will have to push it until it gets a good bite on the end of the pipe. After it begins to work, you will be able to turn the threader without using pressure.

4 · Keep turning the threader until you can see a thread or two on the end of the pipe on the die side of the threader. Then turn the threader counter clockwise to remove it.

5 · Use a cloth to clean the oil and cuttings off the pipe.

WORKING WITH PIPE

SOLDERING PIPE

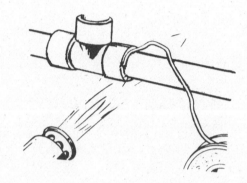

TOOLS

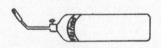
PROPANE TORCH

ADJUSTABLE PIPE WRENCH *Possibly*

MATERIALS

Solder
Flux
Steel wool
Aluminum foil

INSTRUCTIONS

1 · Be sure the inside of the fittings and the end of the pipe are free of burrs.

2 · Polish the end of the pipe with steel wool. Clean the fittings. After cleaning pipe and fittings, do not touch with your fingers.

3 · Put flux on the inside of the fitting and around the end of the pipe.

4 · If you have a threaded pipe, screw it in with a pipe wrench. Otherwise just slip the pipe into the fitting. In either case, wipe out any excess flux.

5 · Shield the area around the fitting with aluminum foil. Check to see that there are no inflammable materials near.

INSTRUCTIONS

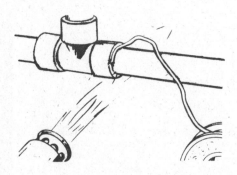

6 · Light the propane torch. Apply the heat evenly all around the joint. Keep the flame moving.

7 · To find out when the pipe is hot enough to solder, occasionally touch the solder to the joint as you heat - touch it on the side opposite the flame. When the solder melts, you can remove the flame and apply all around the joint.

8 · Be careful you don't get the joint too hot or the solder may run all over and you will have to re-clean the joint and start all over again.

9 · The soldering is finished when you can see solder all around the joint.

10 · Allow the solder time to cool before moving it.

11 · If you are going to solder a number of joints on the same line, insure against melting the solder out of a completed joint by wrapping wet rags around the finished joint.

WORKING WITH PIPE

BENDING PIPE: Copper Tubing

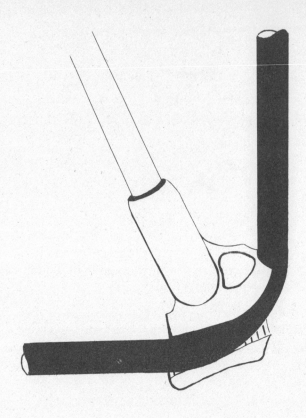

TOOLS

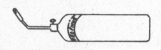
PROPANE TORCH

MATERIALS

Rags

INSTRUCTIONS

There might be a rare occasion when you would prefer to bend the pipe instead of using fittings. This might occur if a pipe came up through floor and is to go into the wall. A baseboard may be in the way, or you may wish to avoid unsightly fittings where they can be easily seen.

Copper tubing

Heat with propane torch where you want to bend it until it appears blue, moving heat back and forth. When bluish in color, with your hands protected from the heat with rags, bend the tubing.

DRAINAGE SYSTEM

The drainage system consists of a series of traps and vented pipes leading either to a cesspool or to a sewer system. Traps are U shaped pipes designed to keep sewer gases from coming up into the pumbing fixtures. Traps are not built to catch tissues, rags or other solids. These will cause fixtures to back up and overflow.

The major problem encountered with the drainage system is clogging. Broken and leaking drain pipes are fixed the same way as water supply pipes.

Waste water passes through the plumbing fixture's stopper and down the drain, where it passes through a trap on basins and tubs. There is usually a clean-out opening on the bottom of each trap. Most fixture drains are vented. For example, a sink will have a re-vent by-pass attached to the main vent. It is necessary to vent the drain so that the rushing water doesn't pull water out of the trap, leaving too little to block off gases.

The waste water passes through the drain pipes to the main drain or soil stack or main stack (all terms meaning the same thing). It goes down the main stack, the upper portion of which is the main vent, and leads to the sewer or cesspool. There may be more than one vent outlet through the roof.

There is a clean-out opening at the base of each main stack for access to blockages. Your system may or may not have a sewer trap, although most areas have ordinances requiring them now. This trap will have clean-out openings on each side - one to clean out the drain leading to the sewer; the other to clean out the drain back to the main stack. You may want to put petroleum jelly on the threads of the plug to make it easier to take off when you need to.

Drain Pipes

Types of drainage pipes that will pass most plumbing codes.

CAST IRON

COPPER TUBING

GALVANIZED IRON

P.V.C. PLASTIC

Note: P.V.C. plastic pipe is so easy to use and light to handle that even a small person could easily use it.

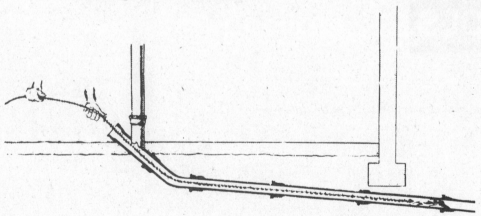

Pitch on drainage pipes to be ¼" to a running foot.

DRAIN PIPE BLOCKAGE

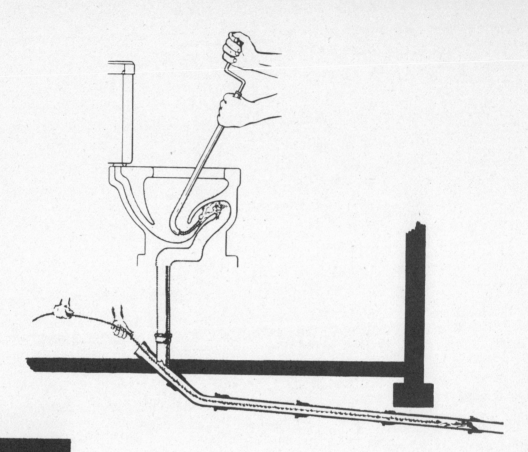

TOOLS

(Tools will depend on each individual job)

MATERIALS

Sewer Tape
Plumbers Helper
Liquid Drain Cleaner
Garden Hose

INSTRUCTIONS

To clean out blockages, you should start at the fixture which is overflowing. If more than one fixture is overflowing, the blockage will be somewhere along the line past where the two or more fixture drains come together. If only one is clogged, follow the steps on this page.

1 · Check the basin or tub stopper. Take it out and clean it.

2 · With stopper out, look into the drain and feel around inside with your fingers for an obstruction. Caution: if the drain opening is larger than your three primary fingers, do not risk getting your fist caught in the drain opening.

3 · If you cannot find an obstruction right near the mouth of the drain, and the water is not completely blocked from draining, use some liquid drain cleaner. NEVER pour liquid drain cleaner into a drain that is completely blocked. Regardless of what the manufacturer says if the product is strong enough to break through a blockage, it is too dangerous to use in a completely blocked drain.

4 · If the drain is completely blocked, or if the liquid drain cleaner was ineffective, use a plumber's helper. Partially fill the basin, tub or bowl, then ROLL the rubber rim of the plumber's helper into the water at an angle, to completely fill the rubber cup with water. Place the rim tightly around the edge of the drain hole. If there is a vent in the basin or tub, place a hand over it - or have someone else do it for you. Give the drain two dozen or more of strong, rhythmic shots. Every few shots lift the cup, from the drain to check for improvement and to let more water into the drain.

5 · If you want to try another way, an ordinary garden hose jammed well into the drain, with rags held around the hose to block the drain opening - and the vent closed off - may work.

6 · If nothing has worked to this point, you will have to use a plumber's snake. The procedure from here on will differ from one fixture to another.

BASIN WITH VISIBLE TRAP
   A. If the trap has a clean-out plug on the bottom, use a monkey wrench to take off the plug. Put a bucket under the trap to catch the water. You may want to put adhesive tape around the plug nut to protect the finish.

   B. If the clog is between the trap and the drain opening, you will be able to tell, because the water will not drain from the basin. You can now use the snake either from the drain down into the trap, or from the trap up into the drain. If the clogging is below the trap, use a 10 foot snake. Feed it in a little at a time. If it gets jammed, pull it out a little and start again. Don't force it. Go slowly, then try again faster. Play around with the blockage when you find it because you may be able to turn any large obstruction to a different angle where it may move more easily.

TUB WITHOUT VISIBLE TRAP

In this case you will have to go in through the drain opening. Again, use a 10 foot snake, using it in the same way as outlined for basin.

INSTRUCTIONS

For the toilet, it is best to use a closet auger, although a regular snake may work too. A toilet has a built-in trap which gives a regular snake a hard time. Toilets are usually built directly adjacent to a main stack so that a 10 foot snake is longer than you need.

MULTIPLE BLOCKAGE or
BLOCKAGE NOT REACHED WITH TEN FOOT SNAKE

1 · If you cannot reach the blockage with a 10 foot snake, or if more than one fixture is blocked at once, you may need to go through one of the clean-out plugs with a 25 or 50 foot sewer tape. This should be rented rather than bought because you would never have to use it often enough to justify the expense.

2 · Introduce the end of the sewer tape through a clean-out opening. If you have a sewer trap with two clean-out openings, open the street side first. Use a monkey wrench to unscrew the end plug.

3 · If the problem is on the sewer side of the trap, waste will be backed up and may make a mess. Be prepared to drain off waste water. It may be best to call a professional sewer man if the blockage is on the street side.

4 · If the street side is clear, and you still have a blockage, open up the house side opening. If the blockage is in the U of the trap itself, there probably will be some waste backed up in sight from grease and soap. Run the sewer tape back up into the waste stack until you find the blockage. Use the sewer tape in the same way described for using a 10 foot snake.

5 · If you do not have a sewer trap within reach, look for a clean-out plug at the base of the waste stack.

OUTDOOR PIPES

If you have an outdoor hose tap or plumbing within the exterior walls of a poorly insulated house, you may run into a problem with frozen pipes. Prevention, as always, is the best cure. Check pipe insulation in your home. If it is in bad shape or non-existent, make it better. During the colder months, turn off the water to all exterior taps.

Winterizing Exterior Taps - Hose Faucets

Every water pipe opening has a shut-off valve within the house somewhere. Trace the location of this valve by starting at the tap and working your way back along the supply pipe.

1 Go inside and turn off the inside shut-off valve to the outside faucet.

2 Leave the outside faucet open all through the winter. In this way, if any water remains inside the pipes, it will be able to expand without bursting the pipes.

Frozen Pipes

If you already have a frozen pipe, you may not be able to tell if your pipe has broken until the water melts. When thawing pipes, be ready to turn off the indoor shut-off valve in a hurry, in case the pipe is broken.

OUTDOOR PIPES

THAWING A FROZEN PIPE

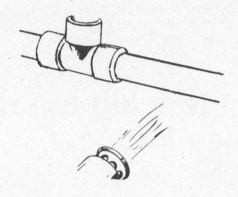

TOOLS

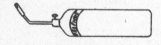
PROPANE TORCH

(This is best, but any heat source will do, even an electric iron)

MATERIALS

Aluminum foil
Lots of rags

INSTRUCTIONS

1 · Shield the area behind the frozen pipe with aluminum foil. Make sure that there is fire stopping equipment nearby if you are using any sort of flame. Beware of gas lines.

2 · Make sure the outdoor tap is wide open.

3 · Make sure the inner shut-off valve is wide open.

4 · Start at the outdoor tap end of the line with your blowtorch or other thawing equipment. Keep your torch moving so that no one spot gets too hot. You should never let the pipe get so hot that you cannot hold it with your bare hands.

5 · Work your way from the tap end back toward the water source. As the ice melts, the water will run out the tap end.

6 · When the pipe is thawed, the water will run freely from the end. At that point, follow the directions for winterizing exterior taps.

NOTE: If the pipe has burst, it will have to be replaced.

INSTALLING SUMP PUMP

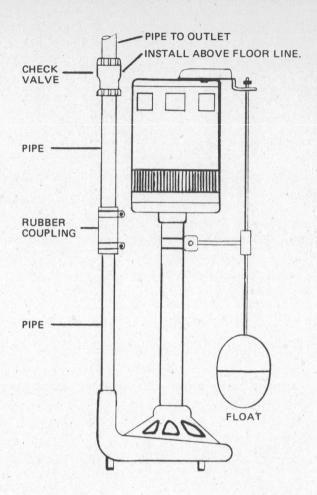

TOOLS

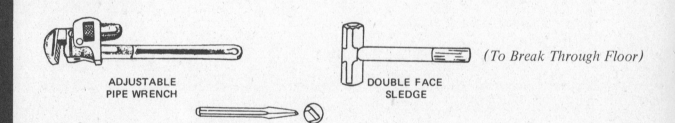

ADJUSTABLE PIPE WRENCH

DOUBLE FACE SLEDGE *(To Break Through Floor)*

COLD CHISEL

MATERIALS

Stone crock *(large enough to set pump in)*
Sump pump
Length of pipe

INSTRUCTIONS

If your home is plagued by periodic basement flooding, you may want to install a sump pump. If the flooding occurs only after unusually severe rain and tide combinations, it might be more economical to rent a small drainage pump, but annual spring basement floods demand a sump pump.

1 · Find where the flooding was the deepest - in other words, where the basement floor is lowest. This is the spot to install the pump. If the lowest point is in an awkward spot, you may want to install a floor drain there with a pipe leading to the sump pump.

2 · Measure the distance from the pump to the nearest exterior wall, or to a place on the sewer line where you can connect.

3 · When buying the pump, discuss the condition with the salesman. He should be able to select a pump which is best for your problem. Find out how high the pump will lift the water before it has to drain. The best pumps are those with a floating control regulator to turn on and shut off the pump.

4 · To install the pump, follow the directions which come with it. You will have to sink it into the basement floor about 18 inches in most cases. Some manufacturers tell you to shore up the sides of the pump hole, but the best bet is to sink some sort of pot and place the pump inside it. The top of the pot should be two inches below the lowest point of the basement floor. As the water level in the ground rises, it will reach the top rim of the pot. As the pot fills, it will start the pump and keep the basement drained.

5 · In setting up the drain lines for the pump, keep in mind the following:
   A. If you are draining out through an exterior wall, be sure the water will drain away from the house foundation. A little extra pipe will take the water far enough away. Otherwise you may find the water circulating senselessly back through the basement.

   B. If you want to connect the sump drain pipe to the sewer line, check local regulations first. It is probably easiest to raise the water and then enter into the waste stack with a Y branch fitting.

A sump pump can be handy for your annual boiler drainage. Maintenance is simple. Merely grease and oil it once a year.